Pi(π), Time and Nikola Tesla 369

Pi(π) Is Time. Lets Do Time Travel

Deepak Singh

Author biography

Deepak Singh is an engineer and a research analyst who lives in India. I'm also happy to call myself an Indian. Deepak Singh is thankful and grateful to All cosmic enregies, gods, parents, friends and nature. In my spare time, Deepak writes and explores space, time, history, and the future. He thinks that information and education can transform the world, and he uses his writing, education, and sharing of ideas to inspire and empower young people. You can find him on the internet at:

https://www.linkedin.com/in/deepak-singh-32a284115/.

CONTENTS

Introduction

For practical computations, ancient civilizations such as the Egyptians and Babylonians need reasonably accurate approximations of. Archimedes, a Greek mathematician, devised an algorithm to approximate with arbitrary precision in 250 BC. Chinese mathematics estimated to seven digits in the 5th century AD, whereas Indian mathematics approximated to five digits, both employing geometrical techniques. A millennium later, when the Madhava–Leibniz series was discovered in Indian mathematics in the 14th century, the first exact formula for was derived, based on infinite series.

It's as simple as counting to one, two, 3.14159265..

If we keep this up, we'll be here for a while. Here's the deal: Pi (π) is the Greek alphabet's 16th letter, and it's used to symbolise the most well-known mathematical constant.

Pi is defined as the ratio of a circle's circumference to its diameter. To put it another way, pi equals circumference divided by diameter (= c/d). The circumference of a circle, on the other hand, is equal to pi times the diameter (c = d). Pi will always work out to be the same number, no matter how big or tiny a circle is. That figure is roughly 3.14, although the situation is a little more convoluted than that. The perimeters of circumscribed and inscribed polygons can be used to calculate Pi(π).

It is simple to decode Nikola Tesla's 369 code, which is a mathematical fingerprint of God. The 369 Code is a technique based on the law of attraction. You may make your wishes come true using this approach, and you can attract anything you desire in your life.

To learn about this approach, we must travel back to the nineteenth century. Nikola Tesla, a great scientist, proposed this notion. However, this technology is not as well-known as Tesla's other inventions since humans only believe in things that are visible and can be applied to our daily lives.

Well, we may not be able to know everything there is to know about the Universe, but we can use its secrets to help us achieve our goals. We must have faith, and only then will we be able to see beyond our limitations. And once you've crossed the finish line, we can achieve any goal we set for ourselves and alter our lives by having faith.

What is the value of Archimedes' constant Pi(π)

Pi is an irrational number, meaning it is a real number that cannot be stated as a fraction. That's because pi is a "infinite decimal," meaning the digits go on forever and ever after the decimal point.

Pi is taught to children as a value of 3.14 or 3.14159 when they first begin learning math. Despite the fact that pi is an irrational number, some people estimate it using reasonable formulations, such as 22/7 of 333/106. (The precision of these rational expressions is only a few decimal places.) Despite the fact that there is no exact value for pi, many mathematicians and math lovers are interested in computing pi to the greatest number of digits possible. Rajveer Meena of India holds the Guinness World Record for memorizing pi to the greatest number of decimal places (while blindfolded) in 2015. Meanwhile, some computer scientists have calculated pi's value to over 22 trillion digits. On Pi Day, a pseudo-holiday that happens every year on March 14 (3/14), calculations like this are frequently revealed.

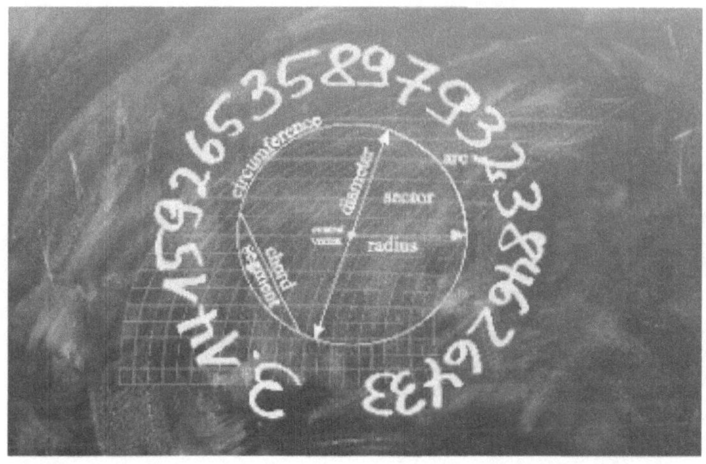

Archimedes, a Greek mathematician, devised a technique to determine the value of pi to 16 decimal places. Since then, pi has been dubbed Archimedes' constant. He drew two polygons, one inside the circle and the other outside the circle, to calculate the value of pi. He repeated the process and discovered that the polygon's sides were growing closer to the form of the circle.

When he made 96-sided polygons, he demonstrated that:
The determination of pi in ancient times was based on measurement. The Egyptians, for example, determined the value of pi (16 / 9)2, which is approximately 3.16. Similarly, while marking the position of the building and limits, the Babylon rope stretchers determined the value of pi 25 / 8 = 3.125.

Because the value of pi does not terminate or repeat, it is an irrational number. As a result, we are unable to calculate the exact value of pi. Pi's decimal equivalent is about 3.14. Pi can also be expressed in fractional form, which is 22/7.

The Life of Pi is a fictionalised account of the life of Pi

Pi has been discovered by ancient Babylonians and has been known for approximately 4,000 years. Pi was found to be 3.125 on a tablet dated between 1900 and 1680 B.C. The Rhind Papyrus from 1650 B.C. shows that the ancient Egyptians were making comparable findings. The Egyptians used a method in this book to compute the area of a circle, giving pi an approximate value of 3.1605. Pi appears to have been approximated in a biblical text as well:

And he fashioned a molten sea, ten cubits from one brim to the other; it was round all around, and his height was five cubits; and it was compassed by a line of thirty cubits. 7:23 (I Kings) (King James Version)

Archimedes of Syracuse was the first person to calculate pi (287-212 B.C.). Archimedes, one of the world's greatest mathematicians, utilised the Pythagorean Theorem to calculate the areas of two polygons. Archimedes calculated the area of a circle using the area of a regular polygon encircled by the circle and the area of a regular polygon circumscribed by the circle. The upper and lower bounds for the size of a circle were determined by Archimedes' mapping of the polygons, and he estimated that pi is between 3 1/7 and 3 10/71.

The British mathematician William Jones first used the pi sign (π) to represent Pi in 1706. Pi was calculated using the formula 3.14159 by Jones.

Pi is used in fundamental mathematics to calculate the area and circumference of a circle. Area is calculated by multiplying the radius squared by pi. So, if you're looking for the area of a circle with a radius of 3 centimetres, 32 = 28.27 cm is the answer. Pi is all around us and is continually used since circles exist naturally in nature and are frequently employed in other mathematical formulae.

Archimedes, Who Was He

In the year 287 BC, Archimedes was born in the city of Syracuse on the island of Sicily. He was the son of Phidias, an astronomer, and mathematician. Apart than that, little is known about Archimedes' childhood or his family. Some claim that he was a member of the Syracuse nobility and that his lineage was related to that of Hiero II, King of Syracuse.

Syracuse was a thriving center of commerce, art, and science in the third century BC. Archimedes acquired his inherent curiosity and aptitude for problem-solving as a child in Syracuse. When he had gained as much as he could from his professors, Archimedes traveled to Egypt in order to study in Alexandria. Alexandria, founded by Alexander the Outstanding in 331 BC, had established a reputation for great wisdom and scholarship by Archimedes' time.

Prior to Archimedes' arrival in Alexandria, Euclid was one of the most well-known scholars in the city. Euclid was a famous mathematician who is most known for compiling all of the existing Greek geometrical treatises into a logical and systematic sequence in his book "The Elements." For nearly 2,000 years, this collection was essential to the study of geometry, and it definitely affected Archimedes' work.

Archimedes returned to Syracuse after his education in Alexandria and followed a life of thinking and innovation. Archimedes endeared himself to King Hiero II by discovering solutions to the king's difficulties, according to many apocryphal accounts.

Archimedes And His Death

Archimedes' genius repelled the Romans for two years, allowing the city to endure the long siege. Nonetheless, Marcellus' armies triumphed and captured the city in 212 BC. Marcellus held Archimedes in high regard and promptly ordered troops to retrieve his foe. Because he was so engrossed in a mathematical problem, it appears that the great mathematician was ignorant that his adversary had besieged the city. When a soldier asked that Archimedes accompany him to Marcellus' quarters, he simply declined and went back to his thoughts. The furious soldier swooped down on Archimedes, killing the quirky 75-year-old. When Marcellus learned of Archimedes' death, he was devastated and ordered that he be buried with honours. After one of Archimedes' geometrical treatises, his tombstone was etched with a depiction of a sphere within a cylinder, as he had requested.

The Legacy of Archimedes

Despite the many fascinating stories about Archimedes' life, we owe him a debt of gratitude for his mathematical treatises and contributions to our understanding of fundamental physical events. He was able to explicate the concepts for such basic devices as the pulley, the fulcrum, and the lever using geometry – technologies that are still used today. The principle of buoyancy, or the ability of a fluid to exert an upward force on a body placed in it, was also discovered by Archimedes. His continued work on volume and density was crucial in the development of theories of hydrostatics, the branch of physics that deals with liquids at rest.

The narrative of Archimedes' treatises making it to our time is intricate and difficult, and it has been meticulously researched. But there is one crucial point: we only know the texts of Archimedes' Greek treatises thanks to three manuscripts. One was last seen in 1311, another in the 1550s, and the third is The Archimedes Palimpsest, which is now on display at The Walters Art Museum in Baltimore and is the focus of this website. Because this is only the beginning of an enthralling tale.

369 Code Key By Nikola Tesla

Nikola Tesla was a remarkable guy and genius who was born to a Serbian family amid a lightning storm in 1856. He had a lifelong fascination for nature and electricity and was resolved to use his gift of enormous mental power for the greater good of society. This "Electrical Wizard with an inexhaustible fire in his blue eyes" wants to know everything about himself and everything around him. He admired women, but he was upset by how they were growing increasingly masculine, competing with men and usurping their rightful place in society. He was well-versed in nutrition and believed in maintaining his body in accordance with natural laws, in tune with the Earth. Nikola Tesla was a firm believer in exercise, "the wonder of milk," and a vegetarian diet, eating meat just a few times a year and preferring fowl. He overcame his addictions by using his willpower; he stopped drinking coffee and tea, which he compares to poison.

He understood that his body was a gift from God:

"Everyone should regard his body as a priceless gift from the one he loves most, a magnificent work of art, of exquisite beauty and mystery beyond human comprehension, and so delicate that a word, a breath, a look, even, a thought, may destroy it."

Because the world was not ready, he was misinterpreted by the world, and his message was never accepted. This renowned physicist and philosopher, who identified as a Christian, illustrated how Science and God are one. He studied the Bible, but his hunger for knowledge also pushed him to read other ancient texts. He was so far ahead of his time because he was open to receiving insights from The Creator Himself, as it is commonly said. He gave his entire life to the betterment of humanity. Radio, electricity (AC), electric motors, X-rays, lasers, wireless communication (internet), and remote control are just a few of the things he "created." Because he knew so much and was prepared to share freely, he posed a significant threat to the world's money-hungry rulers; the FBI invaded his hotel room and stole his paperwork shortly after his death. Nikola Tesla knew a lot more than he was willing to share with the rest of the world. He also understood the value of numbers and that the entire

Universe communicates through numbers. God, the Divine Creator, employs numbers to help us understand His creation; numerology, also known as gematria, is an ancient science that involves equating numbers with letters in order to hide and reveal the hidden meanings of words. In this document, I exclusively used Sumerian Numerology because it is a Sumerian system.

Nikola Tesla aspired to enlighten humankind and considered ignorance as a source of evil; nonetheless, he ran against a stumbling block:

"Of all the frictional resistances, ignorance, which Buddha called "the greatest evil in the world," is the one that slows human progress the most. Only the diffusion of knowledge and the integration of humanity's disparate aspects can alleviate the friction caused by ignorance. There isn't a better way to spend your time.."

Tesla was instrumental in the Creator's Divine design since he focused the world's attention on the numbers 3, 6, and 9:

"You would hold the key to the cosmos if you only knew the magnificence of the 3, 6, and 9."

This trio piqued Tesla's interest, but he also recognized that he couldn't comprehend it:

"My brain is merely a receiver; there is a core in the Universe from which 'We' derive information, strength, and inspiration. I haven't delved into the depths of this core's secrets, but I am aware that it exists."

"Now, I have to tell you about a weird occurrence that had a lasting impact on my life. We got a cold [snap] that was drier than we'd ever seen before. A glowing track was created by people walking in the snow. Maak's back became a sheet of light as I touched it, and my hand generated a shower of sparks. This is nothing but electricity, as my father pointed out, similar to what you see on trees during a storm. My mother seems concerned. She warned him not to play with the cat because he might start a fire. I was pondering something abstract. Is nature a feline? If that's the case, who strokes its back? It must only be God, I reasoned.

The impact of this wonderful sight on my childhood mind cannot be overstated. I kept asking myself, "What is electricity?" day after day, and I couldn't come up with an answer. Eighty years have passed, and I'm still asking the same question, unable to find an answer."

"God, Divine Being, gives us mental power, and if we focus our minds on that reality, we get in sync with this immense force. My mother had taught me to look to the Bible for all the answers."

"... At this point, I did a more thorough study of the Bible and uncovered the key in the book of Revelation..."

The answer to the 369 riddles can be found in Revelation, the Bible's final book:

"Weep not, for the Lion of the tribe of Judah, the Root of David, hath triumphed to open the book, and to free the seven seals thereof," one of the elders says.

5th Revelation

THE JUDAH LION = 888

"And write to the angel of the church in Philadelphia, These things saith he who is holy, he who is true, he who possesses the key of David, he who openeth, and no one shutteth; and shutteth, and no one opens;"

Revelation 3:7 says....

On Christmas Eve 2017, we were asked to walk through

the Tesla 369 code, so I looked at the initial 369 code crack given by Andre Slade on 12/9/2012 (frame below) and realized that it needed to be clarified because it was not straightforward or complete. I'd wanted to do this for a long time since I believed no one fully understood what Andre had said.

The walk-through request was a stimulus, and this post was the reaction.

3: Immaculate conception refers to the conception of the Holy Trinity, Mind, Body, and Spirit, or the waking of the Christos when the spirit comes to life for the first time.

6: The situation: the conclusion of Pisces' gloomy era and the start of Aquarius.

9: The Path: To prepare for what lies ahead, you must balance the 9 enzymes that makeup Oxytocin and detox the body to clear toxins from Laminin (vitamin C = 8 infinite), Proverbs 19: 8 "He who obtains wisdom loves his own soul (body): he who keeps understanding shall find good," Cain (9) signifies a green diet.

If we don't succeed, we'll have to:

3: Individual person: no Immaculate Conception, no Trinity, no Christos Crucifix

6: The situation: Pisces/Satan's circumstances reign supreme and mankind remains exiled.

9: The answer is 25,920 years, which is how long humanity would have to wait for another chance like this. If we don't get it right this time, he may govern once again.

I placed in CAPITALS what YAHWEH revealed to me while working on this explanation; I was looking for the hidden numerical value of words based on what We already knew.

666 CHRISTOS

REBORN (432) ENKI (234) = 666

LAMININ (432) = 666 CLEAR (234)

The Tesla 369 code served as a validation to Andre in its early stages, fuelling His ideology. He had been undertaking extensive research on humankind by that time, and he had discovered some incredible findings; previously published work can attest to this. He didn't realize he was identifying the code of God Almighty Himself at first since he didn't fully appreciate who He was. The code led to His Immaculate Conception (conception of His Spirit) on August 31, 2012, and we now know that no one else could have explained it; it had to come from The Son of God Himself, the one who controls The Universe's Keys.

Connecting the 369 to our cosmic voyage and hence to the current End Times opened the way to ABSOLUTE TRUTH, which was written down in the form of the small book of Revelation 10 titled Where To From Here:

Cognition meant to be conveyed to the world.

COGNITION = 1728 WHERE TO FROM HERE (4 x 432)

"And he said to me, 'You must prophesy before many peoples, countries, tongues, and kings again.'"

10:11 (Revelation)

The number 369 alludes to unity, and hence to YAHWEH, the Almighty Lord.

3: The number 3 symbolizes life's ever-present spiritual component.
The Holy Trinity, which consists of the union of the Neutron (Father), Proton (Son), and Electron, manifests itself in the atom (The Holy Spirit). The Aether is the proton-neutron structure that facilitates electron movement and is the driving force of everything. Elohim's abode, Orion's Belt, also called The Three Sisters, is also represented by the number three.

6: The physical component of life, matter, is represented by the number 6.
Everything has life energy because of the "void," which is the driving force behind the carbon manifestation of matter and energy. Carbon 12 is The Creator's signed building block, with 6 neutrons, 6 protons, and 6 electrons forming the number 666. Six signifies the flesh and earthly belongings, that which is not eternal but is connected to the Earth. It is the union of the mind and the body.

9: The number nine is associated with god consciousness, divine completion, or the all-seeing eye.
It is represented by the pyramid's top and is the result of resonance between 3 and 6.

There is harmony when there is a balance or yin yang between the spiritual and physical lives (resonance). As God gives humankind dominion over all animals, once this divine consciousness develops, it magnifies into divine knowledge with reference to humankind. When the numbers 3 and 6 align, the third eye opens, allowing you to perceive higher dimensions and, eventually, the truth.

The number nine stands for The Way: "Yahshuah says unto him, I am the way, the truth, and the life: no one comes to the Father except by me."

John 14: 6
Anyone who wishes to be reborn must first resonate with King David's Wisdom, which is the Spirit of perfect truth.

Yahshua responded by saying, "Truly, verily, I say unto thee, a man cannot see the kingdom of God until he is born again."
John 3:3

However, not everyone can be reborn, and only a select few will be allowed to enter.
"Verily, verily, I say unto thee, Unless a man be born of water and of the Spirit, he cannot enter the kingdom of God," Yahshua said.

Revelation 3:5

6 6 6 is the result of dividing 3 6 9 into three equal pieces, as in The Holy Trinity.

$$6 + 6 + 6 = 3 + 6 + 9 = 18$$

Life is symbolized by the number 18.

Let me describe resonance as simply and clearly as I can right now: resonance, also known as synchronicity, is two-way cyclic communication between all living things. Resonance is a concept that applies to everything in the Universe and promotes harmony.

It is resonance when you pray to Almighty God YAHWEH and He responds.

It's called resonance when your right hemisphere (brain), which is primarily responsible for the perception and processing of emotions/feelings, sends a message to your left hemisphere (brain), which is responsible for logical, abstract thinking and the development of causal linkages, and the left brain reacts.

It is resonance when your gut-brain (second brain), also known as the enteric nervous system, is healthy because it is kept in balance by beneficial bacteria and sends messages to your main brain (first brain), which then sends signals back to your gut lined with millions of neurons.

When your pituitary gland (body) is healthy and generates oxytocin, it sends and receives messages with your pineal gland (soul), which is known as resonance.

It's resonance when you're traveling overseas and experience acute stress, and your favorite plant at home exhibits stress symptoms at the same time.

It's resonance when a monkey realizes that washing its food (sand and dirt) in water makes it simpler to eat it, and all the monkeys in the world naturally start doing the same.

Everything is made up of light, power, energy, and electricity, and everything vibrates at a certain frequency or the number of vibrations/waves per second.

"Think in terms of energy, frequency, and vibration if you

wish to discover the secrets of the universe." - (Nikola Tesla)

Resonance is the cosmic principle of stimulation and reaction between things of the same sort, and it can resonate or dissonate with humans. Multiples of a given number are linked to it and to one another, resulting in unison.

Earth's creatures benefit from the frequency 432 Hz since it resonates with their bodies. This natural tranquil frequency was modified to 440 Hz, which is energetically destructive. All great ancient civilizations and musicians, whether Sumerians or Egyptians, Mozart, Bach, Chopin, Debussy, or Vivaldi, tuned their instruments to the base frequency of 432 Hz. Stay in balance by listening to 432 music.

The Divine resonance frequency is a combination of 432 Hz and 234 Hz, with 234 Hz symbolizing the Heavens and Above and 432 Hz representing the Earth and Below. The frequency 432 Hz is represented by an upside-down triangle/ pyramid, while the frequency 234 Hz is represented by an upward pointed triangle/ pyramid. The most typical shape of a natural rough diamond is a diamond with two pyramids connecting at their bases when these two opposing frequencies are combined. Tetrahedral units make up the diamond (pyramid with 3 sides and a base). The diamond is the final step of the carbon element and hence represents perfection; the carbon atoms in the diamond are connected in essentially the same way in all directions. This amazing gem is known as ADAMANT in the Hebrew language, and it bears the number 666 in the form of vowels AAA in both Hebrew and Sumerian numerology/ gematria.

God is represented by number 1.
His Two Faithful Witnesses stand behind Him (Zechariah 4:11):
God's right hand is represented by 234 while his left hand is represented by 432:
2341432

To the left side of God, add the right side:

666 = 432 + 234

333 is the midpoint or point of equilibrium between 234 and 432.

Multiply them as follows: 101088 = 234 x 432

These are two of The Three Sisters' characters:

1188 = THE THREE SISTERS

The number 432 can be found everywhere:

30 days or 43,200 minutes make up a biblical month.

There are 43,200 seconds in both the day and the night.

The Annunaki descended from Heaven to rule over the Earth for 432,000 years until Noah's Flood.

The Moon's diameter is 2,160 miles (432: 2 = 216); the Sun's diameter is 864,000 miles (2 x 432 = 864).

As per Hindu Shastra By adding all Yugas = Satya + Treta + Dvapara + Kali = 4320000 years.

According to several experts, the Kali Yuga epoch lasts 432,000 years.

Revelation 13:18 makes a direct reference to King David's wisdom: 234 = 13 x 18

"Wisdom is here. Let anyone who has insight count the beast's number, for it is the number of a man, and his number is six hundred threescore and six."

Satan only stole 666 from Enki, who has the wisdom to unveil the beast.

God's full might is expressed in the number 37, which is the mark of God: 26 + 11 = 37.

666 is the result of multiplying number 37 by powers 3, 6, and 9.

111 = 3 x 37
222 = 6 x 37
333 = 9 x 37
666 = 18 x 37
1 is the result of adding the right and left sides of 37.
You'll get 44 if you add the right side of 37 to it.
You'll get 40 if you add the left side of 37 to it.
When you multiply 3 by 7, you get 21 = 777.
When you multiply 37 by its opposite number, 73, you get 110.
Go from 37 to 73, and you'll get 55 right in the middle.

If you look into the number 37, you will discover that it is absolutely divine. This is Elohim's number. Elohim is the original Hebrew word used in Genesis to describe the Creators of Man; it is a plural form referring to both male and female, and it is none other than Christ, the Son of God, and Sophia, the Holy Spirit. The mitochondrion, which is the powerhouse of YOUR bodies in a microscopic capsule, carries its own circular DNA with 37 genes, which They used to mark Their creation. The Mother gene is the source of all energy. Elohim, or God's corporeal manifestation on Earth, is made up of two entities: YAHSHUAH and His Bride, IMMAYAH. Two merged tetrahedra (triangular pyramids) or the 2-dimensional symbol known as "the Star of David," which signifies 666, are used to depict them. You may also recognize Them as Enki (meaning "Lord of the Earth") and Inanna (meaning "Anu's Beloved One") from history.

666 = ENKI AND INANNA

Andre (Enki) is number three in numerology, while I (Inanna) am number seven, resulting in the number 37, which is characterized as a match made in heaven. The third planet from the Sun and the seventh from Heaven is Earth. The two interlaced triangles ("the Star of David") or pyramids (Merkabah) signify the Union/resonance of Chesed (divine masculine) and Gevurah (divine feminine), YAHWEH's active hands in ruling Earth on His behalf. Andre M. Slade and Katarina Krizani are manifestations of Elohim today. At the age of 37, Andre began his search for the truth in 1999, when he began to see the world in a different light. I flew to South Africa on a one-way ticket on the day Andre announced His Tesla 369 code breakthrough as if it were a divine call. The New Kingdom was founded as a result of our re-union.

KRIZANI AND SLADE = 888
THE NUMBER OF THE NEW KINGDOM IS 888.

Harmony in life can only be attained if divine wisdom is abundant; equilibrium refers to having equal weight on both sides of the scale. It is described in the Bible as the style of life in the New Kingdom:

"And when ye turn to the right hand, and when ye turn to the left, thy ears shall hear a word behind thee, saying, This is the path, walk in it."

30:21 Isaiah

Finding your way back to resonance is not something that happens by accident; it requires humility, perseverance, perseverance, and a lot of research. In the end, Satan has succeeded in destroying God by severing the bond between man and woman.

"The world has seen many tragedies, but the worst tragedy of them, in my opinion, is the current economic situation, in which women compete with males and, in many cases, succeed in usurping men's positions in professions and business. Women's increasing desire to overshadow man is an indication of a decaying civilization." - Nikola Tesla

It's a big difficulty to become one flesh because we're all built as Monopole Magnets that attract our opposites. Andre has a design to Do, and I have a design to Wait (To Be) in The Science of Human Design; he is the single definition, and I am split definition.

Human Design is the only true psychology and the ultimate guide to human connections, with the purpose of removing misunderstandings in interpersonal relationships.

THE HUMAN FORMULA = 888

"Misunderstandings are always the result of a failure to grasp one another's point of view. This is due to ignorance on the part of individuals involved, not so much in their own domains as in their mutual ones." - Nikola Tesla

What did Nikola Tesla have to say about the code?

We know that a circle has 360 degrees when we study circles. We can only get 3, 6, or 9 numbers if we divide a circle several times. We can acquire 90 degrees by cutting a quarter of a circle; that quarter will now be a multiple of 3, 6, and 9. The total degree of the remaining three-quarters angle will be 270 degrees, which is multiples of 3, 6, and 9.

We can see what Nikola Telsa was trying to say with this example. We can see that 3, 6, and 9 play an essential role in our nature in his subsequent innovations. Following the decoding of this pattern, Tesla stated that if you comprehend the power of 3, 6, and 9, you will be able to open the Universe's secret door. After that, your life will be easily changed, and you will be able to obtain anything you desire.

What are the benefits of this strategy for you?

- **Number 3:** In numerology, number 3 is associated with the direct source of the Universe, which is thought to be the source of all human desires. You will be at the base of the Universe if you understand the power of three, and all your wishes will come true. In Hinduism, the number three is considered a lucky number. You may be familiar with the Hindu gods Brahma, Vishnu, and Mahesh, who are known as the Trimurti. You may have also heard of God Shiva's third eye, which represents Truth, Consciousness, and Happiness. Three rivers meet in the Triveni Sangam. We can see the significance of number 3 from all of these facts.

- **Number 6:** The number 6 represents our internal power; by waking this strength, you will be able to flow through the energy, which will act as a means for your wishes to be fulfilled. You will also be able to overcome your impending challenges with this energy. This Universe is thought to have taken God 6 days to create. According to Astrology Venus planet symbolises number 6, which is the sign of love.

- **Number 9:** If our history has had an impact on us, and as a result, we are unable to go forward and accept new things with confidence. Number 9 assists us in separating from our past experiences, as well as reducing self-doubt and negative energy. In order for us to be able to accept new possibilities in the future. According to the Bible, God's pure soul has 9 Fruits of God.

As you can see, the numbers 3, 6, and 9 are highly special, and by using them in a methodical and coordinated manner, we may easily unlock the Universe's hidden door and obtain our desires. However, you must have faith in it.

What should be done in this case?

First and foremost, be clear about your goals. Set realistic goals, such as "I want to be happy and calm," "I want a true life partner," "I want aid to start my business or career," and so on. It will not happen until you put in a lot of effort.

After you've set your goal, envision it for 17 seconds without any distractions, and give it vitality in your mind. Do it four times for a total of 68 seconds (17 + 17 + 17+ 17). Then connect these 17 seconds to the numerals 3, 6, and 9.

Nikola Tesla Who Was He

The Tesla coil, alternating-current (AC) electricity, and the finding of the spinning magnetic field were all innovations of Nikola Tesla.

Nikola Tesla was an engineer and physicist who is most known for inventing the alternating-current (AC) electric system, which is still widely used today. He was also the inventor of the "Tesla coil," which is still utilized in radio technology today.

Tesla was born in modern-day Croatia and immigrated to the United States in 1884, where he briefly collaborated with Thomas Edison before parting ways. He sold various patent rights to George Westinghouse, including those to his AC machinery.

Tesla was born on July 10, 1856, in the Croatian town of Smiljan.

Tesla was the youngest of five siblings, including Dane, Angelina, Milka, and Marica. Tesla's interest in electrical creation was sparked by his mother, Djuka Mandic, who, while her son was growing up, designed modest domestic appliances in her leisure time.

Milutin Tesla, Tesla's father, was a Serbian orthodox priest and writer who pushed for his son to enter the priesthood. Nikola, on the other hand, was mostly interested in science.

During the 1870s, Tesla relocated to Budapest, where he worked at the Central Telephone Exchange after studying at the Realschule Karlstadt (later called the Johann-Rudolph-Glauber Realschule Karlstadt) in Germany, the Polytechnic Institute in Graz, Austria, and the University of Prague.

The idea for the induction motor came to Tesla when he was in Budapest, but after several years of attempting to generate interest in his invention, Tesla chose to leave Europe for America at the age of 28.

What is the relationship between Pi(π), Nikola Tesla's Secret Code 369, and Time

It is correct that $22/7 = 3.142857$ is the value (Approx).

However, if we use the Digit Sum method, we will acquire some information about it.

It's here,

$\pi = 22/7 = 2+2/7 = 4/7 = 0.571428 = 0+5+7+1+4+2+8 = 27 = 2+7 = 9$.

Here we receive the number 9, which is God's mathematical fingerprint.

$\pi = 3.142857 = 3+1+4+2+8+5+7 = 30 = 3+0 = 3$

If this is the case, then $9-3 = 6$ Statement (a)

As a result, we have a value of 3,6,9, which is Nikola Tesla's mathematical fingerprint of god.

We can deduce that there is a link between Nikola Tesla's mathematical fingerprint and Pi(π).

There's a link between Pi(π) and God's mathematical fingerprint. Researchers have discovered that the number 9 is always present in nature and the universe. For the creation, 9 also empowers 3 and 6. 6 contains the 7, 8, and 5 in the same way that 3 holds the 1, 2, and 4, and we have proven by digit sum that = 3 = 0 = 9 Because Pi is linked to God's mathematical fingerprint.

We know that 0 = 9 [Statement 1]

Statement 1 according to Nikola Tesla's mathematical fingerprint of god.
That illustrates the fact that 0 and 9 are both meaningless. However, 9 has value because 0 has worth as well.
Now, we can observe that
π = 22/7 = 2+2/7 = 4/7 = 0.571428 = 0+5+7+1+4+2+8 = 27 = 2+7 = 9 [By Applying Digit Sum Method]

Since we can see from Statement 1 that = 9 = 0, we can conclude that π = 9 = 0.

As a result,
Pi(π) = 0 (Beginning)
Pi(π)= 9 (End)
Pi(π) is now also Time, according to the study. I'd like to point out that Pi(π) also flows across time.
The mathematical fingerprint of god 3,6,9, according to Nikola Tesla, is linked to time.
And, as a result, π = 3 = 9 = 0 flows in Time as well.

Inverse "S" holds white and black energy, according to

Nikola Tesla's mathematical fingerprint of deity and Qi. According to Nikola Tesla's mathematical fingerprint of god 3,6,9, the black dot is 3 and the white dot is 6, and the inverse "S" is where the 9 lies and holds the white and dark energy.

However, because Pi flows everywhere in the cosmos, we can claim that $\pi = 9 = 0$ as a result of Problem Statement 1. Even if we follow all Swastika, there is a 90° angle at each.

What are the nature of every polygon and geometrical shape?

- Nonagon = 1260° = 1+2+6+0 = 9 = π

- Octagon = 1080° = 1+0+8+0 = 9 = π

- Hexagon = 720° = 7+2+0 = 9 = π

- Pentagon 540° = 5+4+0 = 9 = π

- Quadrilaterals (Squares, etc) and Circle 360° = 3+6+0 = 9 = π

- 270° = 2+7+0 = 9 = π

- (Triangles and Semicircle) 180° = 1+8+0 = 9 = π

- 90° = 9+0 = 9 = π

- 45° = 4+5 = 9 = π

- The Area of circle = $\pi r2$

- The Circumference of a circle $= 2\pi r$

- The Surface area of the sphere $= 4\pi r2$

- The Volume of Sphere $= 4/3\pi r3$

- The Surface area of cone $= \pi rl + \pi r2$

- The Volume of cone $= 1/3\ \pi r2\ h$

- The Surface area of cylinder $= 2\pi r2 +h(2\pi r)$

- The Volume of cylinder $= \pi r2\ h$

Note: Pi(π) is in every polygon and almost in every shape of the geometrical figure.

About Qi and Pi: If we write Qi in lowercase as Qi → qi. Now write qi in reverse letter form we see qi → pi. Since pi → π → qi → Qi. In other words, Pi in lowercase as Pi → pi. Write pi in reverse letter form we see pi → qi.
Hence pi → Pi → π → qi → Qi. (Again, directly we can see there is a connection between Pi and Qi)

What is the relationship between Pi(π) and Time?

Let see by Applying Digit Sum,
There are 24 hours in a day. 2+4 = 6
There are 60 Minutes in 1 hour. 6+0 = 6
We know that, 1440 Minutes in 1 Day = 1+4+4+0 = 9 = π
Similarly, In 365 Days (525600 Minutes) = 5+2+5+6+0+0 = 18 = 1+8 = 9 = π
There are 60 secs in a 1 Minute. 6+0 = 6
A Human baby born in 9 Months. 9 = π
Child gets adult at the of 18 years = 1+8 = 9 = π

Also if we calculate the four ages of yugas as give in the Hinduism Rigveda.

Lets see by Applying Digit Sum on ages of all Yugas,

Satya Yuga = 1,728,000 years = 1+7+2+8+0+0+0 = 18 = 1+8 = 9 = π

Treta Yuga = 1,296,000 years = 1+2+9+6+0+0+0 = 18 = 1+8 = 9 = π

Dvapara Yuga = 864,000 years = 8+6+4+0+0+0 = 18 = 1+8 = 9 = π

Kali Yuga = 432,000 years = 4+3+2+0+0+0 = 9 = π

By adding all Yugas = Satya + Treta + Dvapara + Kali = 4320000 years.

Now, 4320000 years = 4+3+2+0+0+0+0 = 9 = π

We Already Found 0 = 9 = π

Since, As Per Hindu Shastra, after Kali Yuga again Satya Yuga will start and so on.

Kali Yuga = 432,000 years = 4+3+2+0+0+0 = 9 = π = 0

After that

Satya Yuga = 1,728,000 years = 1+7+2+8+0+0+0 = 18 = 1+8 = 9 = π

and so on,

That's how Pi(π) flows in time(Past, Present, and Future) from 0(Beginning) to 9(End).

What is the concept of time

Time is a measure of continuous change in our surroundings, typically from a single point of view.

While the idea of time is intuitive and self-evident – the steady passage of events in front of our eyes; the orbit of the Moon around our world – explaining its fundamental existence is far more difficult. Even physicists don't know exactly what happens as time passes. They do, however, have a few theories.

Everyone is familiar with the concept of time, but it is difficult to describe and comprehend. Time is described differently in science, philosophy, religion, and the arts, but the system for measuring it is relatively clear.

Seconds, minutes, and hours are used in clocks. Although the origins of these units have changed over time, they can be traced back to ancient Sumeria. The electronic transformation of the cesium atom defines the modern international unit of time, the second. But what is time, exactly?

Time is described by physicists as the flow of events from the past to the present and into the future. In general, a structure is eternal if it does not change. Time is the fourth dimension of reality, and it is used to describe events that occur in three-dimensional space. We can't see it, feel it, or taste it, but we can monitor its progress.

What is the mechanism of time?

Time has been viewed as a persistent, autonomous force for decades as if the Universe's development were governed by a single clock.

With Albert Einstein's theory of special relativity, this explanation of time changed in 1905.

While the passage of time has long been understood to be linked to space, this groundbreaking theory was the first to unite space and time into a single field of measurements that change based on the relative motion or gravitational forces of objects within it.

To put it another way, time is relative.

Is time a real thing?

Two people traveling at the same speed will accept that their distance and time measurements are identical. However, when one person's pace increases, the other's calculation of time and distance changes as well, even though their own does not.
Time isn't a constant universal unit and there's no need to choose one view of time over another. It's a relative measurement that changes as objects travel faster or slower, or as gravity is increased or decreased.

Gravity bends space-time and slows time: the greater the gravity, the more it bends space-time and slows time. This is why people on the International Space Station, which is further away from Earth's gravity, age significantly slower than those on the ground.

The Time's Arrow

As time moves forward into the future (positive time) or backward into the past (negative time), physics equations function equally well (negative time.) Time in the natural world, on the other hand, has just one direction, which is known as the arrow of time. One of the most important unanswered questions in science is why time is irreversible.

The natural world, for example, follows the laws of thermodynamics. The second law of thermodynamics states that the entropy of a closed system either stays constant or increases. The entropy (degree of disorder) of the universe will never decrease if it is considered a closed system. To put it another way, the universe cannot return to the same state it was in at any stage in the past. Time cannot be rewinded.

Is it possible to turn back the hands of time?

Of course, the shift in speed or gravitational pull must be immense for us to see these results in time. However, as an observer approaches the speed of light, special time metrics become more apparent. Theoretically, when a particle reaches the speed of light, its 'clock' can slow down. From our perspective, once it reaches the speed of light, the clock will appear to be in reverse. Our clock seems to reverse from the particle's perspective.

What about traveling through time?

Similarly, the space-bending volume beyond a black hole's horizon distorts time perspectives.

We have freedom of movement in our Universe, but we are compelled to march in a linear path along the arrow of time. Calculations indicate that crossing the horizon of a black hole swaps certain freedoms. So we wouldn't have to follow time's strict arrow of direction, but we'd lose the ability to move around in space, which would enable us to travel back in time (of sorts).

While these examples help us better understand the essence of time, both light speed and black hole travel have limitations that prohibit us from using them to reverse time in a realistic manner.
At home, don't try either.

Moving forward or backward in time, as opposed to moving between various points in space, is what time travel entails. Jumping forward in time is something that happens all the time in nature. Since Earth moves slower than the International Space Station, astronauts onboard experience a time jump as they return to Earth.

Traveling back in time, on the other hand, has its drawbacks. One problem is causality or the relationship between cause and effect. Time travel can result in a temporal paradox. A classic example is the "grandfather paradox." According to the paradox, you can avoid your own birth if you travel back in time and destroy your grandfather before your mother or father is born. Many scientists claim time to travel to the past is unlikely, but there are workarounds, such as traveling between alternate worlds or branch points, to a temporal paradox.

What is Time Dilation and How Does It Work?

Time is the same everywhere in classical mechanics. The synchronized clocks are still in sync. However, time is subjective, as we know from Einstein's special and general relativity. It is dependent on the observer's frame of reference. This can trigger time dilation, in which the time between events lengthens (dilates) as one gets closer to the speed of light. Moving clocks are slower than stationary clocks, and the difference becomes more noticeable as the moving clock reaches light speed. Time is recorded more slowly in jets or orbit than on Earth, muon particles decay more slowly as they fall, and the Michelson-Morley experiment confirmed length contraction and time dilation.

Perception of Time

The human brain is capable of keeping track of time. Regular or circadian rhythms are regulated by the suprachiasmatic nuclei of the brain. However, time perception is influenced by neurotransmitters and medications. Chemicals that stimulate neurons to fire more rapidly than normal speed up time perception, while reduced neuron firing slows it down. As time seems to pick up, the brain recognizes further events within a given interval. When one is having a good time, time really does appear to pass.

During emergencies or times of risk, time seems to slow down. The brain does not speed up, according to researchers at Baylor College of Medicine in Houston, but the amygdala becomes more active. The amygdala is the memory-making area of the brain. Time seems to be slowing down as more memories are created.

The same phenomenon explains why older people expect time to pass more quickly than younger people. According to psychologists, the brain develops more memories of new experiences than of old ones. Time seems to move more quickly later in life because fewer new memories are formed.

Time's Beginnings and Ends

Time had a beginning in the world, as far as we can tell.

The Big Bang happened 13.799 billion years ago, marking the beginning of time. We can calculate cosmic background radiation as microwaves from the Big Bang, but no earlier radiation exists. One reason for the beginning of time is that if you could go back in time indefinitely, the night sky will be illuminated by older stars.

Will the time come to an end? This question's answer is uncertain. Time would continue if the universe continued to extend indefinitely. Our timeline would end and a new one would begin if a new Big Bang occurred. Random particles emerge from a vacuum in particle physics experiments, but it's unlikely that the universe will become static or eternal. Only time will tell if this is true.

Models of space-time can define changes in time and space from one point to the next, but they don't explain anything about time's insistence on following a set of events.

Our Universe is a single block of space-time according to these time definitions. There's a point – the Big Bang – before which we can't apply our best interpretation of the laws of physics. There comes a point at which change is no longer meaningfully calculated. However, no single moment in time stands out as visually distinct as 'now.' "Physicists like us know that the distinction between past, present, and future is nothing more than a stubbornly enduring illusion," Einstein once wrote.

However, there may be several clues to the mystery of time in fields other than cosmology. For example, Austrian physicist Ludwig Boltzmann suggested a relation between time and the levels of disorder in the Universe in the 1870s.

It pointed at a potential explanation for why time's arrow points forward by tying thermodynamics' theory of entropy to time that only travels in one direction: maybe our Universe evolves from a low entropy, highly compact infant Universe to a highly disordered, vast Universe drifting into the future.

What exactly is a Yuga

The is divided into four "Yugas" – ages, epochs, or periods of time – each lasting tens of thousands of human years, according to Hindu philosophy. The Satya Yuga, the Treta Yuga, the Dvapara Yuga, and the Kali Yuga are the four yugas. According to Hindu cosmology, the Universe is formed in its entirety once every 4.1 to 8.2 billion years, only to be destroyed completely once again. For Lord Brahma, the Creator of the Universe, this is thought to be one complete day and night. The lifespan of a Brahma is estimated to be between 311 trillion and 40 billion years. These Yugas, like the waxing and waning of the moon, the four seasons, and the rising and ebbing of tides, are thought to repeat themselves in cyclical cycles.

Each of these four Yugas includes phases of transformation, evolution, and metamorphosis, in which not only the physical world but also mankind's entire thought process and consciousness metamorphoses for the better or for the worse, depending on the Yuga. A Yuga's entire cycle begins at its pinnacle, the Golden Age of Enlightenment. From there, it progresses in stages until it enters a Dark Age of evil and ignorance, after which it returns to the Golden Age to complete the cycle. Hindus conclude that one Yuga cycle represents the time it takes for the solar system to revolve around another star.

Each of the Four Yugas has a different length of time.

The length of each Yuga is as follows, according to the Laws of Manu, which was the earliest known text describing the four yugas in detail:

12,000 years equals 4800 years + 3600 years + 2400 years + 1200 years. Since this number represents just half of a loop, the whole cycle will take 24,000 years to complete. This is also an example of equinox precession.

The exact length of a year of life for demigods is not specified here. However, according to the most recent interpretation of the Shrimad Bhagavatam, the Satya Yuga lasts around 4,800 years, which corresponds to the period of the demigods. The Dvapara Yuga lasts approximately 2,400 years, while the Kali Yuga lasts approximately 1,200 years. As a result, one could possibly deduce from these figures that a demigod's year would be roughly 360 human years.

This will also lead us to conclude that the Satya Yuga lasted approximately 1,728,000 years or 4,800360 years. The Treta Yuga, on the other hand, lasted 1,296,000 years and lasted 3,600 360 years. Likewise, the Dvapara Yuga lasted 2,400 360 years or 864,000 years. The Kali Yuga is said to be the shortest of them all, living just 1,200 360 years, or 432,000 years in total. The four Yugas pursue a 4:3:2:1 timeline ratio, as can be seen from the aforementioned figure.

As previously stated, humanity as a whole is witnessing a steady decline in wisdom, intelligence, intellect, life span, physical, and spiritual strength with each passing generation. This means that dharma, or justice, is in decline and is being destroyed.

- Satya Yuga: Dharma reigned supreme during the Satya Yuga, with human stature measured at 21 cubits. At the time, the estimated human lifespan was 100,000 years.

- Treta Yuga: In this Yuga, virtue fell to a fifth of what it was in the previous one. The estimated human lifespan was 10,000 years, and human height was valued at 14 cubits.

- Dvapara Yuga: Virtue and sin were split into equal halves in the Dvapara Yuga. The average human height was 7 cubits, and the average human lifespan was 1000 years.

- Kali Yuga: The Kali Yuga has only a quarter of its time dedicated to virtue, with the rest devoted to sin. The average human lifespan is about 100 years, and human height is reduced to 3.5 cubits. It is estimated that by the end of this dreadful Dark Age, the total human lifespan will be reduced to 20 years.

The Satya Yuga:

The Satya Yuga is the oldest and most important of the four epochs. On Sunday, Vaishakh Shukla Tritiya, also known as Akshaya Tritiya, this period began. This can go back as far as 17, 28,000 years. In this period, God took four different forms: Matsya, Kurma, Varaha, and Narsimha. In this period, knowledge, reflection, and penance will be especially important. People were taller on average back then than they are now. Every king will achieve the predetermined goals and experience happiness. All four pillars of religion were present in their entirety: reality, penance, yagna (religious sacrifice), and charity. Manu's Dharma Shastra was the only text that was considered credible and followed. After the Kali Yuga, Kalki will re-establish the Satya Yuga.

The Satya Yuga will begin at the end of this period when the Sun, Moon, and Jupiter all join Pushya Nakshtra, the Cancer Zodiac. The stars/constellations will become auspicious and radiant during this period. As a result, all creatures' well-being will improve, and their fitness will improve. Kalki, Vishnu's Incarnation, will be born into a Brahmin family at this auspicious moment. Following this, all future generations will adhere to Bhagwan Kalki's values and participate in religious activities. As a result, when the Satya Yuga begins, all will be assiduously engaged in good, sublime deeds.

Beautiful parks, Dharmasthanas (Resting Inns), and magnificent temples will spring up around you. Countless massive yagyas will be performed. Brahmins, sages, and ascetics would be immersed in penance according to their nature. The evil and deceivers will not be found in ashrams. This age would usher in improved agriculture, with the ability to produce all food crops in all seasons. People would contribute generously and adhere to all of the rules and regulations. The kings must zealously protect their people and the earth.

The Treta Yuga

In the Treta Yuga, religious sacrifices, or Yagyas, become common. One of religion's four pillars has come to an end. People at this age will be honest and will carry out all religious rituals in accordance with the sacrifices. It is during the Treta Yuga that overtures of Yagyas, faith, and related practices can be seen. People will reap desired results by performing deeds, making Vedic contributions, and making resolutions.

Everyone in this period was diligent and active. Truth, that is, honest speech, good behavior, and respect for all creatures, was the main religion of Brahmins. Yagyas, self-study, and donation were the common religion of all Brahmins. Shudras' primary goal was to serve Brahmins, Kshatriyas, and Vaishyas. The security of people, agriculture, commerce, and poultry were the responsibilities of Kshatriyas and Vaishyas, respectively. All would truly carry out their responsibilities, and they would be rewarded with heavenly happiness as a result.

In the Treta Yuga, the average human life expectancy was about 3000 years. All Kshatriyas born during this time were brave, zealous, big thinkers, pious, honest, lovely, and deserving of being blessed, respected, and protectors of all people.

The Dwapar Yuga

In the Dwapar Yuga, religion has been reduced to only two pillars. Penance and charity were the only things on people's minds. They were kingly and in search of pleasure. Since divine intelligence had vanished during this period, few people could be trusted to say the truth. As a result, people were afflicted with illnesses, diseases, and a variety of desires. People will perform penance after suffering from these ailments. Some people would hold a yagna for both material and spiritual reasons.

The Kshatriyas of this period were modest and regulated their senses to perform their duties. The king would seek the guidance of learned scholars and, as a result, maintain law and order in his kingdom. The king who was addicted to vices would undoubtedly lose. The kings were meticulous in upholding public order and decorum.

Many a conspiracy was planned behind the scenes by kings and scholars. When it came to enforcing laws, strong people will be the ones to do the job. The king will select priests and other religious figures to carry out religious practices, economists and ministers to carry out monetary activities, the powerless to look after women, and cruel men to carry out heinous acts.

Penance, faith, sense of power, discipline, yagna, and other practices will help Brahmins achieve celestial bliss. Charity and hospitality will help Vaishyas ascend to higher planes. The Kshatriyas will carry out all policies of law and order honestly, without being angry, cruel, or greedy, and thus achieve bliss. By nature, everybody in this period was zealous, valiant, courageous, and competitive.

The Kali Yuga

Kali Yuga's length and chronological start point in human history have resulted in a variety of evaluations and interpretations. Kali Yuga started at midnight (00:00) on 18 February 3102 BCE in the proleptic Julian calendar, or January 3102 BC in the proleptic Gregorian calendar, according to the Surya Siddhanta. Many Hindus believe that this date marks the departure of Krishna from Earth.

Hindus claim that during the Kali Yuga, also known as the Dark Age, human society spiritually degenerates because people are as far removed from God as they can be. Morality (dharma) is often depicted as a bull in Hinduism. The bull has four legs in the Satya Yuga, the first stage of creation, but morality is reduced by one-quarter in each generation. Morality has been reduced to a quarter of what it was in the golden age by the time of Kali, and the

Dharma bull has only one leg.

Yugas' references in the Mahabharata

At the Yuga-Sandhi, the point of change from one Yuga to the next, the Mahabharata War and the decimation of the Yadavas occurred. According to the scriptures, when Duryodhana was about to be born, Sage Narada intercepted the demon Kali on his way to the Earth in order to make him an incarnation of 'arishadvargas' and adharma in preparation for the period of merit decay and the ensuing havoc.

The majority of Hindu scripture interpreters agree that the Earth is now in the Kali Yuga.

What is Chi and why do we need it

Universal Energy, essential life force energy, or simply energy is how "Chi/Qi/Ki" is described. Chi is a Chinese philosophy that dates back to ancient times. In some Asian nations, such as Japan, it is known as "ki," while in Korea, it is known as "gi." The word's spelling can vary depending on the community, but the context remains the same. Chi is the movement of energy that affects all living things, including trees, plants, humans, and animals. The ultimate concept practised in ancient Chinese traditional medicine and Martial Arts is essentially synonymous with the term "life force." The literal sense of the word "qi" or "chi" is "breath" or "air." Another interpretation of the word "qi" or "chi" is "breath" or "air."

Introduction of Chi

The origins of chi can be traced back to China, with evidence of its presence in ancient texts dating back to the 5th century BC and closely linked to the Chinese religion Taoism. Although there is no evidence of this term in Western cultures, it is known in Hinduism as 'prana,' which is a Sanskrit translation of life force, and its Hawaiian equivalent is known as 'mana.' It is referred to as 'ki' in Japan and as 'chai' in Hebrew.

The metaphysical element of the core definition of chi has been a source of controversy in China, as there are many interpretations. In China, Taoists and Buddhists claimed that chi is the source of all things, while others believed that chi can come from any physical matter and that it is a different force from the physical world. All in the universe is made up of energy that vibrates at various frequencies. Energy flows into living beings not only through physical nourishment but also through meridians and chakras, which are energy centres. Chi (Qi or Ki) is the force of life itself, a balance of Yin and Yang, positive and negative, electromagnetic energy that pervades all of existence. As a result, Chi may be classified as an electromagnetic phenomenon, a type of light energy, a type of bio-electromagnetic energy, or electricity.

Chi's Principles

The definition of chi is that it is the fundamental component of all. Chi offers energy or strength in the same way as a fresh breath of air provides life to all living things. Qi, also known as chi, is the primary cause of human life and a factor that influences the human body's health. It is assumed that each of us has a balanced chi balance and that the harmonious flow of chi determines our fitness. It is beneficial not only to one's physical health but also to one's mental faculties. The pleasure of being alive, or the breath of creation, is the life force energy known as chi.

Chi's Mysteries

"Steam rising from rice as it cooks," as described in ancient Chinese writings on chi, is quite similar to this idea. Steamed rice is China's main staple food, and the perfect balance of water, which is considered "yin," and fire, which is considered "yang," is needed to cook the rice grains. This thinking reveals the secret of chi theory, which calls for an exact combination of yin and yang as a symbol of nourishment in order to live a long life. Because of the rising steam, some assume Chi stands for oxygen, which better explains how chi feels in the body as motion or floating.

Chi Kung (Qigong) literally translates to "energy work" or "energy cultivation." There is work to be done in order to become physically stronger, emotionally linked, mentally concentrated, spiritually aligned, and energetically vitalized. Work on the physical, emotional, mental, spiritual, and energetic levels that will enable us to develop and manifest a healthy physical body, a pure caring heart, and an open creative mind.

Keeping the Chi in Perfect Balance

Chi, as respected by the Chinese, denotes a balanced lifestyle, both psychologically and physically, and can be improved through the practice of ancient Chinese breath control exercises known as Tai'-Chi, as well as acupuncture and calming massage. Almost all of the methods used in Chinese medicine are strictly based on Chi principles. Traditional Chinese medicine practitioners believe that strengthening one's chi, or qi, is associated with increased overall energy and good health. Chi Kung/Qigong is a Chinese martial art that incorporates meditative based breathing exercises, movement, visualisation, and purpose to encourage wellbeing, healing, peace, wellness, and inner joy. Yoga, Chi Kung, Tai Chi, Healing Sounds, and Meditative Techniques all aid in the cultivation and purification of chi energy.

We must become a complex and harmonious combination of all facets of chi in order to become a joyful, balanced, and happy human being. Chi is in a constant state of transformation, shifting from one element of chi to the next. It is neither formed nor destroyed; rather, it varies in its appearance. Chi can be developed and harnessed for our benefit by using various chi kung and meridian stretching methods, gentle movements, breathing exercises, and meditation.

The Human Body and Chi

The human body, according to traditional Chinese medicine, has chi that flows across the system in a continuous pattern. Meridians are chi flow patterns like this. The meridians are thought to be accessible via the fascia, which is a connective tissue underneath the skin. Unbalanced chi flow in the body can cause a variety of health problems and illnesses, as well as affect a person's psychological well-being. Any chi imbalance in the body can affect the organs, exposing the body to diseases and systemic disorders. Acupuncturists use needles to help unblock the flow of chi energy, which can cause pain and illness. Many different types of chi have been defined by Chi Kung practitioners. This is our original chi, essence chi, or jing chi, which we inherited and are born within our bodies. It is the root cause of metabolism and the human body's fundamental chi. This allows us to lay solid foundations, gain root stability, and gain support for ourselves, allowing us to more easily cultivate and retain chi energy and, as a result, become a powerful medium for Spiritual Light.

The Art and Science of Living are these ancient Yogic and Taoist methods. We can become physically energised, emotionally linked, mentally concentrated, and spiritually aligned through practice. We may learn to identify, control, and mitigate stress, which has naturally occurring beneficial side effects such as increased self-esteem, trust, self-discipline, and personal well-being. The act of balancing the chi in the body through traditional Chinese therapeutic approaches and other practices will help fix the imbalances and lead to better health and longevity. There are a variety of methods that can help you keep your chi in check. Qi Gong, Tai Chi, herbal treatments, acupressure, acupuncture, physiotherapy, and meditation are examples of these practises. The lower dan tien is its true home. The primordial feminine and masculine powers, Yin chi and Yang chi, are the two most basic kinds of chi.

QI Gong, Tai Chi, and other martial arts are also widely recognised as beneficial to health and well-being around the world. Moving, breathing, and mental awakening are all used in these ancient arts to align and stabilise our bodies and minds. To experience the energy flow of chi, Qi Gong masters must practise complex movements for years. To keep the energy flow inside my body balanced and allow me to experience the Qi Gong's flow of chi, I personally use a Chi Machine on a daily basis. Our Chi Kung practises emphasise the relation to and use of Heaven and Earth chi, as well as chi that comes from Nature, such as trees, rivers, seas, lakes, and mountains. Heaven Chi is inhaled with each breath.

Meat, water, and air absorb earth chi, and chi kung/qigong practise can be referred to as post-natal chi. The combination of prenatal original chi and postnatal chi is the body's true chi or complete energy. In the middle dantian, we can cultivate purified chi. The heart channel supports the pulse and voice by promoting chi and blood circulation and by surfacing in the throat. Protective chi is the chi that flows at the surface of the body and protects us. Spleen chi, lung chi, kidney chi, heart chi, stomach chi, and other internal organs have chi.

Lung chi descends and Liver chi ascends, for example, balancing each other. When you're upset, your liver chi rises, causing tension in your shoulders and neck, tightening of your jaw and facial muscles, chest constriction, and an urge to protest and scream. You can allow the Lung chi to descend harmonising and calming it all down by letting it out constructively, using healing vibrations, or calming yourself down by deep breathing. Healthy chi is harmonious, balanced, and free-flowing, while the disease is caused by stagnant, blocked, or imbalanced chi.

Finally, we will learn how to be happy (no matter what), foster Inner Joy, and Build Personal Resilience. Tai Chi, Chi Kung (Qigong), Healing Sounds, Meditation, and

Yoga are wholesome disciplines for perfect life harmony, where all our joy, happiness, fitness, abundance, compassion, and love can be found.

They are fantastically strong tools for living radiantly in this universe, but they also serve a far higher purpose: they can help us bridge the distance between heaven and earth. We can change our lives forever with practice, cultivating a strong balanced physical body, harmonious positive feelings, a pure caring heart, and an open creative mind, as well as cultivating an unbreakably strong link to our Higher Selves, Divine Love, and Universal Energy.

Trimurti, who are they

The Hindu Trinity, also known as the Trimurti (meaning "three forms of God"), is an iconographic depiction of God in Hinduism that portrays divinity as a three-faced deity. These three faces reflect God's positions of creation, preservation, and destruction, which are identified with Brahma, the creator, Vishnu, the preserver, and Shiva, the destroyer and transformer, respectively. These three personae are said to represent various manifestations of the one supreme divinity. In this way, the Trimurti resembles certain Christian Trinity interpretations, such as Sabellianism. However, these parallels should not be taken too far. Beyond the superficial three-in-one similarity, Christian trinitarianism differs from "Trimurti" in almost every way, with the exception of Shankara's interpretive scheme from the ninth century.

This Trimurti idea is most firmly held in the Hindu Smartism denomination, though it is generally opposed by other Hindu denominations including Saivism and Vaishnavism.

Trimurti's Evolution

The Hindu gods Brahma, Vishnu, and Shiva (who together form the Trimurti) each have their own mythology, scripture, and folklore sources. However, how they came together in a single iconographic representation remains a topic of debate among academics. The Trimurti's origins can be traced back to the Rigveda, where the earliest expression of god in three ways can be found, according to scholars. The all-important aspect of fire is represented in three ways in this work: Agni in the hearth, Vidyut as lightning, and Surya as the sun. Meditation on the One shows it to be represented in a set of triadic deities, one of which is the triad of Gods Brahma, Rudra, and Vishnu, according to the Maitrayaniya Upanishad 4.5.

Shiva and Vishnu had ascended to the top of the Hindu pantheon by the time the Hindu Epics were written (500-

100 B.C.E.). Attempts were made in the Epics to associate Shiva with Agni, the god of fire who is revered in the Vedas. In one passage of the Mahabharata, for example, Brahmins declared Agni to be Shiva. In the case of Vishnu, he already had a position in Vedic mythology, and he was sometimes granted dominance as a supreme personal Deity. His well-known presence as Krishna in the Bhagavadgita only added to his illustrious reputation. The three gods as modes of one greater being, on the other hand, play almost no part in the Epics. The concept of Trimurti is only stated in the appendix of this work (10660 ff). Brahma, on the other hand, is generally overlooked, while Vishnu and Shiva are regarded as equal parts of an androgynous force known as Hari-Hara.

The Trimurti did not become a normative doctrine until the advent of the Puranas, a vast corpus of mythical and historical Hindu texts. The origin of the three modalities of the one supreme Vishnu is explained in the Padma-Purana, a Vaishnava text: "In order to shape this universe, the supreme spirit created from his right hand." Brahma is a Hindu god. He produced Vishnu from his left side in order to keep the universe running. To kill it, he birthed Shiva from his womb. Some men worship Brahma, while others worship Vishnu, and still, others worship Shiva. The pious should not distinguish between these three because they are one." This is the first time the three gods' fundamental oneness as constituents of the supreme concept has been mentioned explicitly. It should be remembered, however, that the trinity was never actually worshipped.

Around the second millennium of the common period, iconographic depictions of the Trimurti first appear. The famous image of the Trimurti statue, for example, was carved between the eighth and tenth centuries on Elephanta Island (near Mumbai, India). The imperial Rashtrakutas of Manyakheta (in modern-day Karnataka), who ruled the southern and central parts of India during this time period and used the image of the three Gods as their insignia, are credited with creating this sculpture. Shiva is represented in this illustration as manifesting all three facets of the Godhead.

The advent of the Trimurti has been suggested as a deliberate attempt to reconcile the major Hindu deities of the time into one universal Godhead in order to reduce divine rivalry and foster unity and harmony among devotees. The Trimurti, like the Hindu deity Harihara, represents the deep impetus in Hindu thought towards inclusion and syncretism from this viewpoint. Vishnu, Shiva, and, to a lesser degree, Brahma were known by various names depending on the location in which they were worshipped prior to the existence of the trinity. Via traditional poetry and art, among other mediums, they eventually came to subsume the names and characteristics of deities with whom they shared a common spirit. Consider some of Vishnu's alternate names, such as Vasudeva and Vaikuntha, for example. It's possible that Vishnu is confused with Indra, another Vedic deity. As these gods rose to prominence in various regions' common beliefs, their characteristics became synchronised with the powers attributed to Brahma, and they came to embody the Supreme Personal Being in their own right. As a result,

Brahma, Vishnu, and Shiva were chosen to reflect a triple Godhead as it manifests in the existence, preservation, and destruction of the universe, respectively.

The Trimurti's Three Divinities

The Trimurti's three forms or faces reflect God's functions of creation, preservation, and destruction, which are identified with Brahma (the source or creator), Vishnu (the preserver or indwelling-life), and Shiva (the destroyer and transformer). Some Hindus combine the three gods' cosmological functions to form the acronym "GOD," which stands for Generator (Brahma), Operator (Vishnu), and Destroyer (Vishnu) (Shiva).

Brahma is a Hindu god:

Hindu gods are often depicted with specific icons and animal companions or "vehicles" in iconography. The Swan is Brahma's vehicle. Brahma is usually painted red to represent the sun's creative force. Four heads, four ears, and four arms define his physiognomy. According to Hindu mythology, he had five heads at one time, but Shiva cut off the fifth to keep Brahma from falling in love with Shatarup, a female deity. One of the four Vedas is recited by each of Brahma's remaining heads. He is normally portrayed with a white beard, suggesting his elder god status. Brahm is the Lord of Sacrifice since one of his four hands is seen holding a sceptre in the shape of a spoon, which is connected with the pouring of holy ghee or oil into a sacrificial pyre. Brahm keeps a mala (a string of rosary-like beads) in the other side, which he uses to keep track of the universe's length. He is also depicted holding the Vedas and, on occasion, a lotus flower. A water-pot is held by a fourth hand (sometimes depicted as a coconut shell containing water). He is said to reside in Brahmapura, a legendary city on Mount Meru. Before the great Shakti supplanted Brahma as the creative force of divinity, Brahma reflected the creative power of divinity (feminine Goddess). Brahma was self-born (without a mother) inside the lotus that emerged from Vishnu's navel at the beginning of the world, according to the Puranas. According to some legends, Brahm is the offspring of Brahman, the Supreme Being, and Maya, his female energy. According to another legend, Brahm created himself by first making water and then depositing his seed in it, which developed into a golden egg. Brahma was born as Hiranyagarbha from this golden egg, and he is also known as Kanja (or "born in water"). The remaining materials of this golden egg are said to have spread into the Universe. Brahma created ten Prajapatis during evolution, according to another part of Brahma's mythology (Fathers of the human race, as well as seven great sages). Saraswati, the goddess of literacy, peace, and creative endeavour, is often seen with Brahma.

Vishnu is a Hindu god:

Vishnu represents the supreme divinity's active, loving hand. Vishnu, Hindus believe, incarnates on a regular basis to preserve righteousness (dharma) and destroy evil, and he is most notably associated with his avatars, especially Krishna and Rama. His name literally means "all-pervading one," and it is thought to originate from the tale of his three-stride measurement of the universe, as told in the Rig Veda's 'Vishnu Sukta.' After beating Indra in mythological mythology, Vishnu rose to dominance in the Hindu pantheon. Since the lotus that spawned Brahma and then the universe arose from Vishnu's navel, the Visvakarma Sukta of the Rig Veda (10.82), which tells the tale of Brahma's existence, appears to refer to Vishnu indirectly as the Supreme God. This storey was reinterpreted in the Puranas to imply that Brahma merely imagined himself to be the firstborn, and that true authority over life belongs to Vishnu. The great Hindu Epics depict Vishnu's ascension to dominance. The Bhagavadgita, a part of the larger Mahabharata Epic, is perhaps the most important example of Vishnu's power. In this case, he appears as Krishna, the charioteer for Arjuna, a conflicted warrior. Arjuna is urged by Vishnu's avatar to pursue the direction of righteousness and duty rather than selfish desires.

Vishnu is usually portrayed as a man with four arms. He's all-powerful and all-pervasive essence is symbolised by his four arms. He is often portrayed holding four talismanic objects: a conch shell (whose sound reflects the primordial sound of creation), a chakra, and a talisman (a discus-like weapon that symbolises the mind without ego), a lotus flower and a Gada (a mace from which mental and physical power is derived) (or the Padma, which represents liberation through dharma). Vishnu is normally depicted in blue, which reflects the sky and the ocean's pervasiveness. He wears the auspicious "Kaustubha" jewel and a flower garland around his neck. His head is adorned with a crown, symbolising his absolute authority. Vishnu is depicted with an earring in each ear, symbolising the intrinsic opposites in life, such as wisdom and ignorance, happiness and unhappiness, and so on. Sri or Lakshmi, the goddess of beauty and good fortune, is Vishnu's consort.

Vaishnavites (those who worship Vishnu as the supreme deity) often believe that nothing is actually destroyed and that Shiva's destructive force is merely a transformation of matter. As a result, the matter is never truly destroyed, and Vishnu's ability to preserve life is declared to be the ultimate force in the universe. In addition, he is revered in the form of his avatars. Vaishnavism is very common in modern India, particularly in the north, and it has also spread beyond India in the form of Hindu diaspora and Gaudiya Vaishnavism, which came to North America in the 1960s through the International Society for Krishna Consciousness (ISKCON).

Shiva is a Hindu god:

Shiva is the personification of the ultimate divinity's destructive force. Shiva's predecessor Rudra, the Vedic god of death and the wastelands, appears to have inherited this destructive aspect. Including the fact that he is referred to as an annihilator, Shiva is regarded as a positive force since existence inevitably follows destruction, and creation at new and higher levels is contingent on Shiva's annihilation's cleansing force. Many Shaivites (Shiva devotees) have nuanced the conventional concept of Shiva in this way. recasting him as the personification of God's reproductive capacity, a view that appears to have supplanted Brahma's status as the founder. Shiva, according to devotees, is more than a destroyer; he also serves as a creator, preserver, and destroyer, as well as bestowing blessings on worshipers. Shaivites speak of Shiva in the same way as Vaishnavas speak of Vishnu as the Ultimate Truth. Shiva is portrayed in Shavite mythology as the being who reconciles all polarities observed in the physical world, based on his power to both kill and build. Shiva is thus both static and active, as well as the oldest and youngest, virile and celibate, gentle and fierce, and so on. He also reconciles men and women's duality by taking the form of Ardhanarishwara ("half woman, half man") to declare men and women's equality. He is often said to be omnipresent, dwelling as pure consciousness in every living being.

Shiva, like Vishnu, is not bound by personal traits and is capable of transcending all qualities and iconographic representations. In light of this, Hindus often depict and worship Shiva in symbolic forms, such as the Shiva linga (or lingam), a phallic clay mound or pillar with three horizontal stripes. Shiva is often depicted in deep meditation atop Mount Kailash, Shiva's traditional abode in Tibet's south. Shiva's body is covered in cemetery ashes, symbolising that death is the true truth of creation. As a result, Shiva is typically depicted in white. The third eye on his forehead reflects his ability to see beyond the obvious, as well as his uncontrollable energy that kills evildoers and their sins. Shiva also wears the crescent of the fifth day (Panchami) moon on his head. This reflects the sacrificial offering's influence as well as his mastery of time. Shiva wears tiger, deer, and elephant skins to reflect his mastery of desire, pride, and the mind. Shiva also wears a poisonous cobra around his neck to show that he has defeated death. Shiva wields a trident, a weapon that is used to punish evildoers on the divine, subtle, and physical worlds. The three prongs also reflect the divine triad's artistic, preservative, and destructive roles, while Shiva's Trident affirms that all three aspects are essentially under his power. Shiva and his consort Parvati (also known as Shakti) are inseparable since they are considered one in the absolute state of being. In his form as Ardhanarishwara, Shiva is said to share half of his body with Shakti. Shaivism is the most widely practised branch of Hinduism in South India today.

The Importance of Meaning:

The Trimurti has been interpreted in a variety of ways, particularly in terms of cosmology. The three gods seen together are thought to embody earth, water, and fire, according to popular belief. The earth is known as Brahma because it is seen as the source of all creation. Vishnu is the god of water, which is seen as the sustainer of life. Shiva is known to be fire since it absorbs or transforms life. Alternatively, the Trimurti's three members are thought to embody the three planes of consciousness: on the celestial level, Brahma represents the spiritual aspect, Vishnu represents the psychic element, and Shiva represents the physical element. Brahma represents intuitive and imaginative thinking, Vishnu represents wisdom, and Shiva represents emotion on the psychic level. The sky is Brahma, the Sun is Vishnu, and the Moon is Shiva on the physical plane. The Trimurti is said to reflect different stages of a person's life. Brahma represents the first of these stages, celibacy and studentship (Brahmacharya Ashram). Awareness, as embodied by

Brahma's consort Saraswati, is the individual's constant companion during this period. Vishnu represents the second step of adulthood and householders (Grihastha Ashram). During this process, the person fulfils all religious and familial obligations by working to generate money, which is then used to support the family. Wealth is the individual's companion during this time, and it is portrayed by Vishnu's consort, Goddess Lakshmi. Shiva represents the third step, which is old age (Vanaprastha Ashram). The renunciation of the material world for an austere existence devoted to the pursuit of true knowledge marks the beginning of this period. This was traditionally the time when a householder, along with his companion, left their worldly possessions to live in the forest with only their basic belongings, as Lord Shiva did. The person seeks union with the Supreme Power in the final process (Sanyasa Ashram) (Isvara). In the same way that the Trimurti culminates in a transcendent One, the three phases of life culminate in a transcendent One, bolstering the belief that the three gods are one and the same Isvara.

Another cosmological interpretation of Trimurti was given by the philosopher Shankara (c. 788–820 C.E.). Shiva describes the Nirguna Brahman (or Brahman without features), Vishnu the Saguna Brahman (or Brahman with features), and Brahma the Cosmic Mind, according to him. In a more metaphysical sense, Brahma is linked to Divinity's Creative Ground of Being, while Vishnu is linked to Divinity's Emanated Idea (Logos, Wisdom, or Word), and Shiva is linked to Divinity's Transformative Energy (Flame, Breath, or Spirit).

Who Were the Ancient Astronauts Called the Anunnaki

The popularity of all types of media relating to the mythology of the ancient Mesopotamians has skyrocketed in the modern period. The writings of a number of researchers who suggest links between many Sumerian myth cycles and the hypothesis that the human race was orchestrated or formed by a group of extraterrestrial beings are fueling this ever-growing trend.

This area, known as Ancient Astronaut Theory, is largely based on translations of cuneiform tablets allegedly created by Zecharia Sitchin, whose Earth Chronicles books serve as the basis for the modern church of the alien gods.

The Anunnaki, according to Sitchin, are a race of mythical beings who mixed their DNA with that of Homo erectus to establish mankind, with the intention of using humans as slaves to mine gold and other minerals. Today, the Anunnaki are often depicted as the Old Testament's creator God. But, in terms of the Anunnaki and other mythic beings, what does the cuneiform corpus really say? How does the Ancient Astronaut media portray these beings and their behaviors compare to how they were actually depicted in the ancient world?

Bloodlines of Prince:

For starters, Anunnaki means "princely blood" or "seed of Anu," not "those who came down" or "those who came from heaven to earth," as many modern sources say. The Anunnaki are a pantheon of gods who were the children of the sky god Anu and his sister, Ki, and are described as "the Sumerian deities of the old primordial period." Some scholars have concluded that the Anunnaki should be classified as demi-gods or semi-divine beings, rather than gods. Ki, Anu's sister, was apparently not initially considered a deity and was only given the title of goddess far later in the myth cycle's history.

According to William Klauser:

"Some scholars doubt that Ki was regarded as a god since there is no evidence of a cult and the name occurs in only a few Sumerian creation texts. Samuel Noah Kramer connects Ki to the Sumerian mother goddess Ninhursag, claiming that the two were once one and the same. She went on to become the Babylonian and Akkadian goddess Antu, consort of the Sumerian god Anu.

This would imply that the Anunnaki were created through the union of a sky god and a mortal woman, who was later deified in mythic traditions.

From the Dust to the Dust:

Furthermore, the Sumerian symbol for "planet" is "Ki," and Anu's consort is often thought to be the personification of the earth. This is analogous to Biblical legend, in which mortals were made from the earth's dust (Genesis 2:7). The idea of a community of half-divine beings born of mortal women is very close to the Biblical and non-Biblical Nephilim tradition.

The Extra-Biblical Book of 1 Enoch, attributed to the patriarch Enoch, son of Jared and father of Methuselah, is one of the most heavily cited ancient texts that describe the Nephilim. Today, 1 Enoch is considered an apocryphal text and most modern theological establishments condemn it. This, however, was not always the case. Many early Church Fathers acknowledged the book as scripture, including Athenagoras, Clement of Alexandria, Irenaeus, and Tertullian, and fragments of ten copies of 1 Enoch in

Aramaic have been discovered among the Dead Sea Scrolls. 1 Enoch is also mentioned in Jude, a Biblical book. There are several hundred more parallels in the New Testament itself, according to some estimates.

Daughters of Man and Sons of God:

The most well-known parts of 1 Enoch contain an explanation of some events that occurred prior to the biblical flood (specifically Genesis chapter 6, verses 1-4). According to 1 Enoch, a party of 200 fallen angels known as the Watchers descended upon Mount Hermon, led by a man called Semyaza (or Semjaza), where they took an oath to father human women's lineages Both of these "took wives for themselves, and each chose one for himself, and they began to go in unto them and defile themselves with them," resulting in the birth of "great giants" as a result of their marriage.

These giants finally "consumed all of man's acquisitions" and "turned against men and devoured mankind" when "men could no longer support them." 1 Enoch, Chapters 6-7) God responds by cursing the giants to wage war against one another as a result of their actions. "so that they will kill each other in battle," and the archangels are sent to bind the Watcher leadership "in the valleys of the earth." (Enoch 10:1) The Hebrew texts refer to the strong beings born to the Watchers as The Nephilim, as is well known today.

Facts

- We may conclude that by using digit sum Pi(π) value is from the beginning to the end time based on research, facts, and the use of maths (Yugas). π = 0 = 9.

- Pi(π) has a link to a newly born human infant (baby takes 9 months to born).

- Each polygon and geometrical 2D and 3D figures are connected to Pi(π).

- Pi(π) is used in nature.

- Pi(π) is connected to the mathematical fingerprint of god 3,6,9, which was discovered by Nikola Tesla.

- Pi(π) will take us on a journey through time.

- We can see this, pi → π → qi → Qi.

- Archimedes made a trip to Egypt. The great pyramids of Giza are well-known in Egypt. The triangle is represented by the pyramid form. 369 also represents a triangle, By the Digit Sum Method (180 Degrees = 1+8+0 = 9).

- On March 14, 1988, physicist Larry Shaw created Pi Day at San Francisco's Exploratorium Science Museum. He's also known as "Pi's Price." The day of constant pi is commemorated every year. The 14th of March was chosen because it is the first three digits of pi (3.14). It is for this purpose that we commemorate Pi Day on March 14th.